SCIENTIFIC AMERICAN EDUCATIONAL PUBLISHING

SCIENTIFIC AMERICAN INVESTIGATES THE HUMAN BODY

THE RESPIRATORY SYSTEM

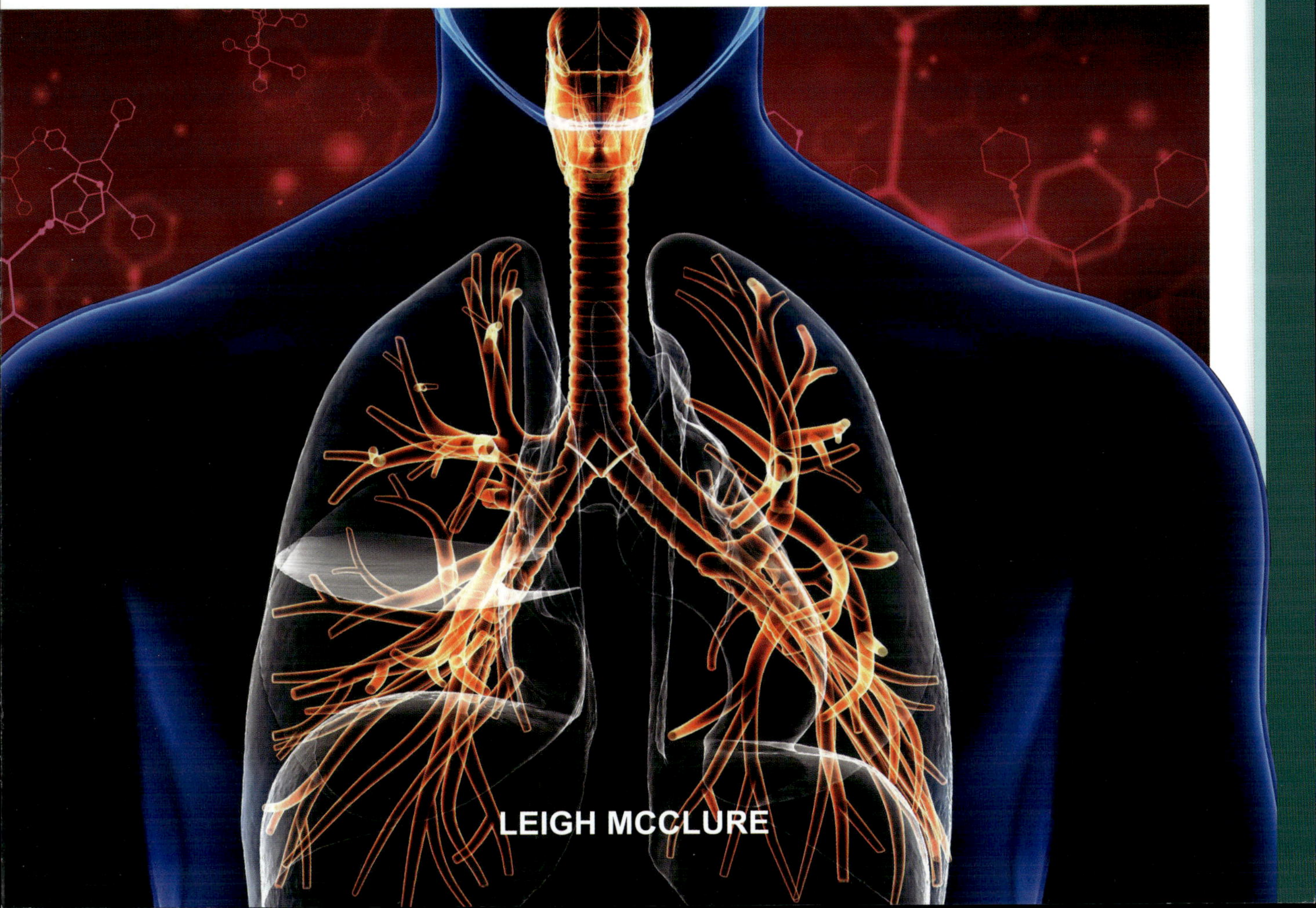

LEIGH MCCLURE

Published in 2025 by The Rosen Publishing Group
in association with Scientific American Educational Publishing
2544 Clinton Street, Buffalo NY 14224

Cataloging-in-Publication Data
Names: McClure, Leigh.
Title: The respiratory system / Leigh McClure.
Description: New York : Scientific American Educational Publishing, an imprint of Rosen Publishing, 2025. | Series: Scientific American Investigates the Human Body | Includes glossary and index.
Identifiers: ISBN 9781725351479 (pbk.) | ISBN 9781725351486 (library bound) | ISBN 9781725351493 (ebook)
Subjects: LCSH: Respiratory organs–Juvenile literature.
Classification: LCC QM251.M34 2025 | DDC 611'.2–dc23

Designer: Andrea Davison-Bartolotta
Editor: Caitie McAneney

Portions of this work were originally authored by John M. Shea, MD, and published as *The Respiratory System*. All new material in this edition authored by Leigh McClure.

Photo credits: Cover, p. 1 (main) Life science/Shutterstock.com; cover, p. 1 (background) jijomathaidesigners/Shutterstock.com; p. 5 wavebreakmedia/Shutterstock.com; p. 7 PeopleImages.com - Yuri A/Shutterstock.com; p. 9 VectorMine/Shutterstock.com; p. 11 Africa Studio/Shutterstock.com; p. 13 (main) Liya Graphics/Shutterstock.com; p. 13 (inset) first vector trend/Shutterstock.com; p. 15 Meeko Media/Shutterstock.com; p. 17 Fit Ztudio/Shutterstock.com; p. 19 Gorodenkoff/Shutterstock.com; p. 21 Ground Picture/Shutterstock.com; p. 23 Mita Stock Images/Shutterstock.com; p. 25 Studio Romantic/Shutterstock.com; p. 27 Dragana Gordic/Shutterstock.com; p. 28 matimix/Shutterstock.com; p. 29 Pressmaster/Shutterstock.com.

Printed in the United States of America

CPSIA compliance information: Batch #CSSA25. For Further Information contact Rosen Publishing at 1-800-237-9932.

CONTENTS

Words in the glossary appear in **bold** type the first time they are used in the text.

BREATHE IN, BREATHE OUT

Breathing is **essential** to life. Humans breathe about 15 times a minute. That's more than 20,000 times a day! However, we often aren't aware of this basic body function. Whether at rest or playing around, your body needs oxygen. This gas helps turn food into energy. We breathe in, and our lungs fill with oxygen-rich air. We breathe out, and air containing a waste product called carbon dioxide is released from the body.

We breathe in and out because of the respiratory system. The organs that make up this system work each day without you thinking about them so that you can have all the oxygen you need. When one part doesn't work correctly, it can cause some big problems. It's time to explore this breath-taking body system!

In yoga, people practice becoming aware of the breath. Some even practice breathing exercises, such as making their exhale longer than their inhale.

What Is Air?

We breathe air each day. But what's in it? About 21 percent of air is made up of the gas oxygen, which our bodies use to produce energy from food. Nitrogen gas makes up 78 percent. The remaining 1 percent contains small amounts of gases such as helium, methane, and water vapor. Carbon dioxide gas makes up only 0.04 percent of air. Most scientists think that even a small increase in carbon dioxide can make temperatures on Earth hotter.

ESSENTIAL GASES

Living things need oxygen. Without it, we'd die! Animals need oxygen to transform foods, such as sugar and fat, into energy. For **bacteria** and other small living things, oxygen simply crosses **cell membranes** to get where it's needed. For larger animals, such as humans, a **circulatory system** delivers oxygen throughout the body. The job of the respiratory system is to provide the circulatory system with a continuous supply of oxygen.

When cells use oxygen and sugar to make energy, carbon dioxide is produced as waste. Too much carbon dioxide can be poisonous to animals. So, another job of the circulatory system is to help move carbon dioxide from the cells to the lungs. The lungs work to remove carbon dioxide from the body.

Plants not only provide food for us to eat, they help keep our air clean and fresh as well.

The Role of Plants

Plants play a vital role in our environment. During photosynthesis, plants use energy from sunlight to change carbon dioxide into sugar. In doing so, plants produce oxygen as a waste product and release it into the air. The flow of oxygen and carbon dioxide between plants and animals shows how different organisms, or living things, depend on each other. Animals depend on the oxygen plants produce, while plants use the carbon dioxide released by animals to make food.

THE MOUTH AND NOSE

The mouth and nose are part of the upper respiratory tract. That's where oxygen comes into the body. Although we sometimes breathe through our mouths (for example, when we're talking), we most often breathe through our noses. The inside of your nose is lined with many tiny hairs and a layer of **mucus**. These act as a filter that prevents dust, bacteria, and other dangerous microscopic particles from traveling deep into your lungs where they may cause illness or injury. The body tries to clear dust and other particles filtered by the nose by exhaling very forcefully: a sneeze!

The mouth and the nasal passage join in an area called the pharynx, or throat. The throat divides into two separate tubes: the airway, which leads to the lungs, and the esophagus, which leads to the stomach.

That Smells!

Every time we inhale, air flows over many specialized cells in our nose called odor receptors. These cells are very sensitive to many different chemicals in the air and can recognize over a trillion unique odors! Our sense of smell can act as an early warning for danger, such as the smell of smoke or the odor of rotten food before we eat. Some animals, such as dogs, have a much greater sense of smell than humans.

FUN FACT

AT THE BOTTOM OF THE THROAT IS A FLAP OF **CARTILAGE** CALLED THE EPIGLOTTIS. IT GUARDS THE AIRWAY FROM FOOD THAT HAS BEEN SWALLOWED.

Breathing and Swallowing

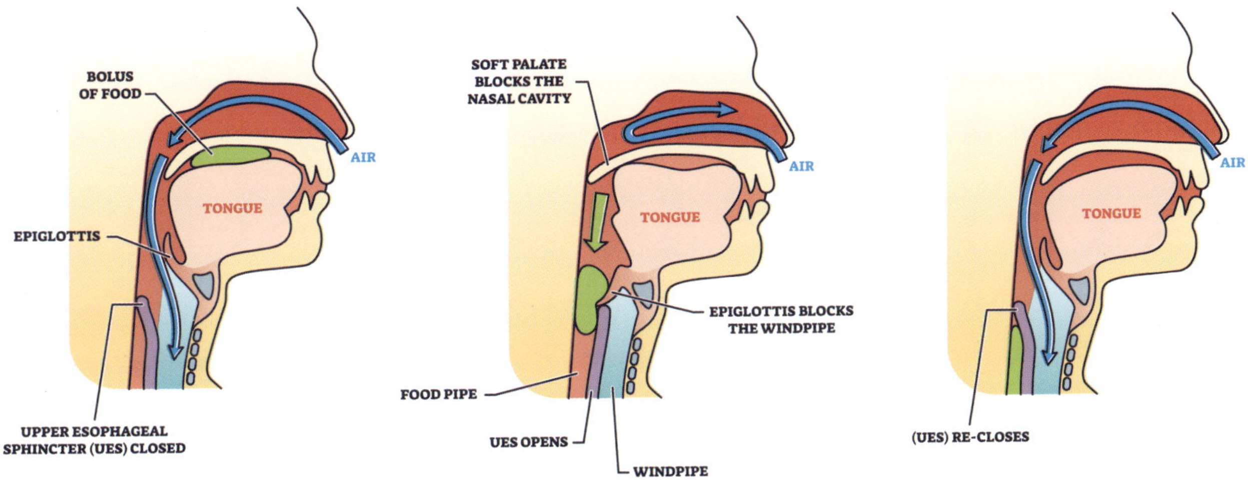

Check out this diagram to understand how the movement of air and food works in the upper respiratory tract.

SPEAK UP

The trachea is also called the windpipe. It leads from the throat to the middle of the chest. To prevent this hollow tube from collapsing, it's held open by cartilage rings. Above the trachea, just below the epiglottis, is a special tube of cartilage known as the larynx, which contains the vocal cords.

During normal breathing, the vocal cords are relaxed and held open so air may pass through them. When we speak, the cords tighten and vibrate, producing sounds. The more closely the cords are drawn together, the higher the pitch of the sound they make. Adult men tend to have a larger larynx than women, which creates a deeper voice. A larger larynx results in a laryngeal prominence, or bump, that is called the "Adam's apple."

FUN FACT

THE TRACHEA IS A TUBE THAT'S ABOUT AS WIDE AS A FINGER.

The sound of our voice is further changed by how our lips and tongue are shaped while air escapes through our mouth.

When You Choke

Although the epiglottis closes the airway when we swallow, food particles sometimes enter the trachea and block air from flowing. This is called choking. Choking can be deadly when the victim can't get enough air. Some people hit the choking victim on the back to try to knock the food out. However, this can make the food go deeper into the lungs. Instead, a method called the Heimlich maneuver uses air in the victim's lungs to force the food up and out of the trachea.

A LOOK AT THE LUNGS

The trachea divides into two branches in the middle of the chest. Each branch, or bronchus, enters a lung and divides into smaller tubes, called bronchioles. Each bronchiole divides into branches that become smaller and smaller until they're thinner than a human hair.

At the very end of each tiny bronchiole are extremely tiny sacs called alveoli. The walls of alveoli are so thin that oxygen can pass through them and flow into the nearby **capillaries**, beginning its journey through the circulatory system. Likewise, carbon dioxide in the capillaries flows through the alveoli walls to enter the lungs so it can be exhaled. An average adult has about 300 million alveoli. If you could stretch them out flat, they would cover a tennis court!

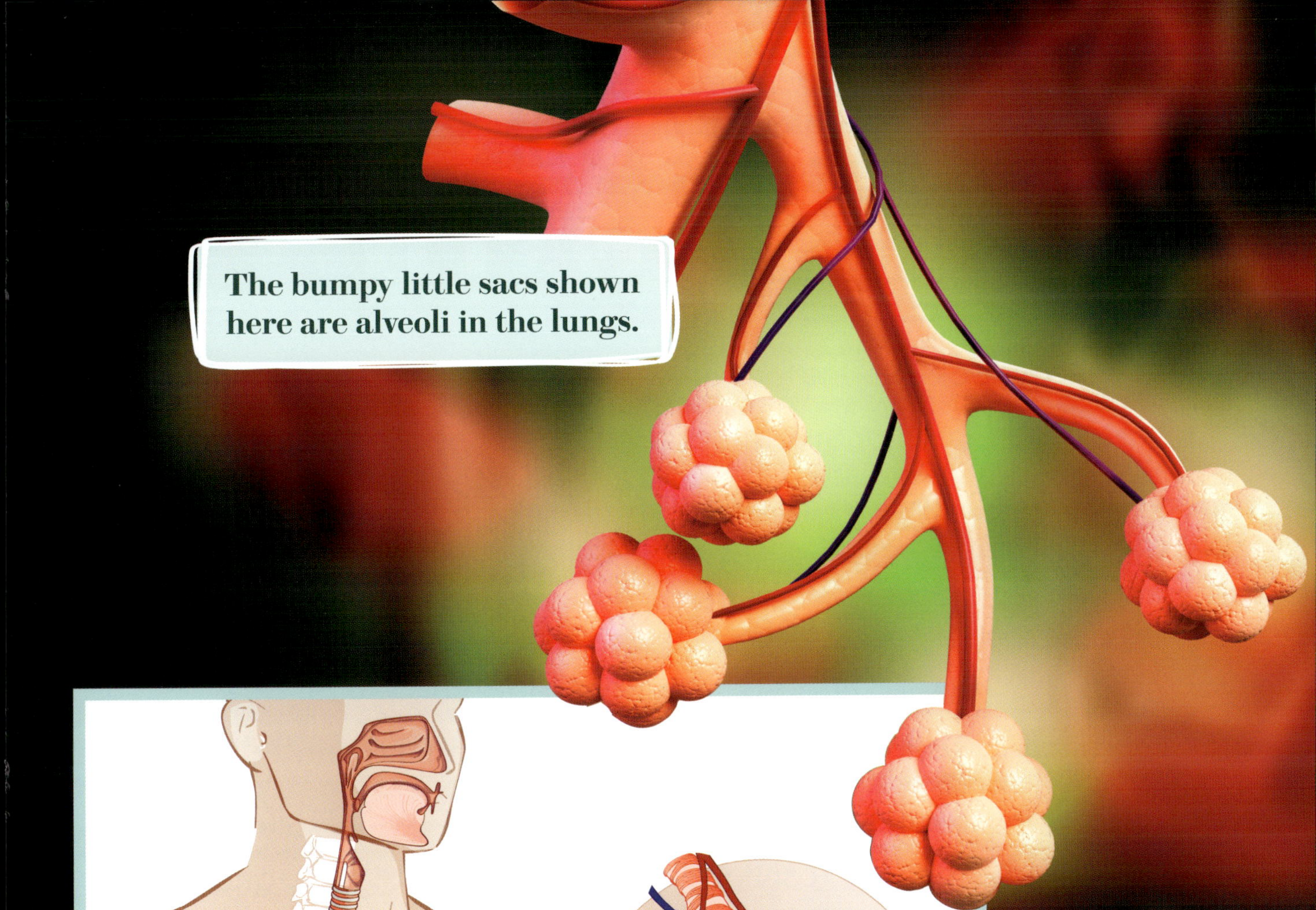

The bumpy little sacs shown here are alveoli in the lungs.

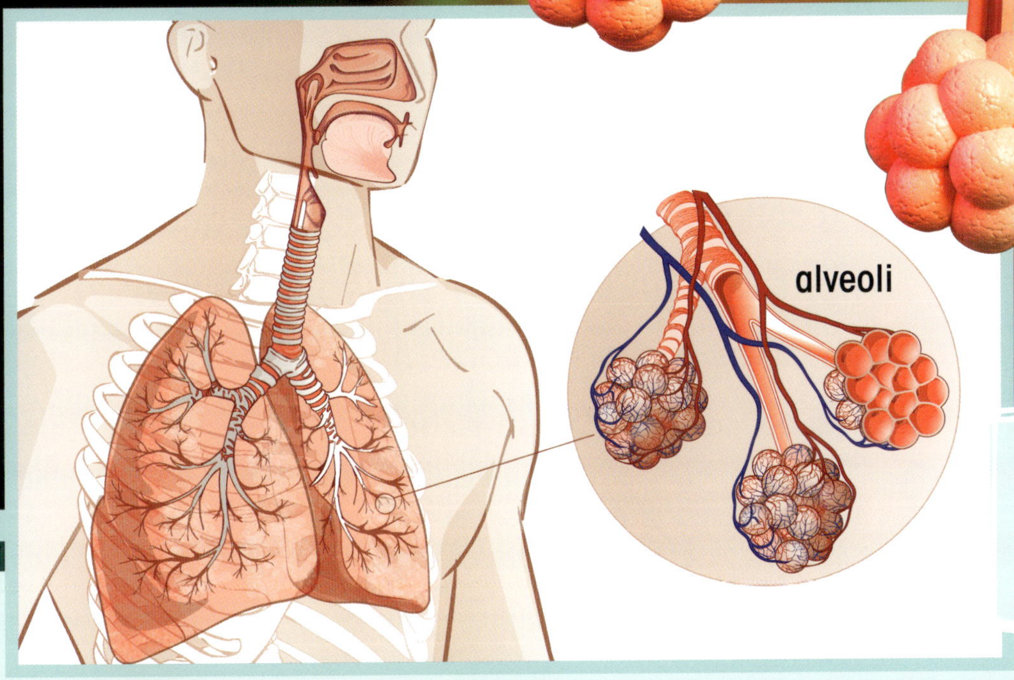

Check out what alveoli look like at the end of each bronchiole.

THE SCIENCE OF BREATH

Let's look into the path that a breath takes. When we breathe, we're changing the size of our lungs by contracting and relaxing the muscles in our rib cage and our diaphragm. The diaphragm is a sheet of muscles stretching across the inside of our body just below the lungs.

First, we take in breath and our diaphragm contracts, which stretches the lungs down. At the same time, the muscles of the rib cage contract, which stretches the lungs wider. These actions result in the lungs getting bigger, and air is sucked into them. That's an inhale. Then, we relax the diaphragm and the muscles of the rib cage. This squeezes the lungs, which forces air out. That's an exhale.

Put one hand on your belly and one hand on your chest. Breathe in and out, noticing where and how the breath is moving.

On Autopilot

There's a specialized area of the brain, called the brain stem, that controls many of the body's automatic functions. These include the beating of the heart and breathing. Special sensors in the body, called chemoreceptors, monitor the levels of oxygen and carbon dioxide. If the chemoreceptors sense that more oxygen is needed—for example, when we exercise—the brain stem signals the diaphragm to contract more quickly.

VIRUSES AND INFECTIONS

Most people know the basics of how to prevent illness. You can wash your hands often and stay away from people who are sick. Children routinely receive vaccines for numerous illnesses, including mumps, polio, measles, and chickenpox. However, it's still normal to get **viruses** and infections. That's because the respiratory system is exposed to the outside environment. It's constantly in danger of catching something, especially in the winter months. The common cold is one of the most frequent human illnesses.

Influenza, or "the flu," is also caused by viruses, but it can be more serious. Many of us recover from influenza within a week. However, it can be deadly for people who have weak **immune systems**. COVID-19, for example, can be serious for those with weak immune systems.

Very old people and very young babies are at a higher risk of serious illness from viruses. Make sure to keep your distance from them if you're sick.

FUN FACT

COLD VIRUSES ATTACK AND KILL THE CELLS IN OUR NASAL PASSAGES, CAUSING SYMPTOMS SUCH AS RUNNY NOSE, SORE THROAT, AND COUGHING.

Shots Save Lives

Vaccines (sometimes called "shots") can sound scary, but they save lives! When you get a vaccine shot, a tiny amount of a dead or weakened virus or bacteria enters your body. These germs can't hurt you, but they do prompt your body to build a defense. Then, when your body is exposed to the strong, living germ, it's ready to fight. You won't get as sick, and you're less likely to spread the disease to anyone else!

Some people are prone to getting pneumonia, which is a serious lung infection. It can be caused by viruses, bacteria, or funguses. While these germs are in the lungs, they damage the thin, delicate walls of the bronchioles and alveoli. When this happens, those parts can no longer supply oxygen to the blood. In addition, the thin walls of the alveoli allow these germs to enter the bloodstream, which spreads them throughout the body.

Without treatment, pneumonia will worsen. The lack of oxygen can become life threatening. **Antibiotics** and other medications can help fight the infection and help the patient breathe better. Some cases of pneumonia can be treated at home. Serious cases require the constant attention of trained medical professionals in a hospital.

Some pneumonia patients need higher amounts of oxygen in the hospital until their lung function improves.

Cleaning Crew

When you inhale, you breathe in many tiny particles, including dust, pollen from plants, and bacteria. Alveoli contain specialized cells called macrophages, which come from the Greek words for "big eaters." The macrophages do exactly what their name suggests: they literally eat all the bacteria, dust, and other particles they can find in the alveoli. This helps keep the lungs clean and healthy.

UNDERSTANDING ASTHMA

Imagine not being able to breathe enough to play sports or run around with friends. Some kids with asthma face this issue often. During an asthma attack, the bronchi in the lungs begin to swell, squeezing the space inside them so that less air can pass through. People with asthma must work their breathing muscles extra hard to keep air moving into and out of the lungs. They often make a wheezing sound as they exhale. There's always a danger that an attack lasts so long that the patient becomes too tired to breathe.

Fortunately, there are good medications available to both treat and prevent asthma attacks. Albuterol is an inhaled medication that quickly widens the bronchi walls. Some children may need to keep their inhaler close by if they want to play sports.

FUN FACT

ABOUT 6.5 PERCENT OF CHILDREN IN THE UNITED STATES HAVE ASTHMA.

Some asthma patients use a nebulizer, which turns medication into a mist that they can inhale.

Asthma Triggers

In the majority of asthma cases, the swelling of the bronchi is triggered by some external factor. In many cases, it may be cigarette smoke, dust, pet dander, pollen, or cold air. In some cases, the asthma attack occurs when the person has an infection. It's very important that people with asthma and their families and friends do their best to avoid these triggers to help prevent attacks.

DISEASES AND DISORDERS

Sometimes the alveoli become permanently damaged and scarred. This leads to a disease called emphysema, which is a form of chronic obstructive **pulmonary** disease, or COPD. Air exchange becomes extremely limited, and patients are short of breath all the time. Emphysema is a progressive disease, which means patients usually get sicker, not better.

Bronchitis is the irritation and swelling of the bronchi. Similar to what happens with asthma, the swelling makes it more difficult to move air through the bronchi, and the patient may feel short of breath. In this case, the swelling occurs over a much longer period of time. Bacteria, viruses, and cigarette smoke are all common causes of bronchitis. Patients usually have a cough that can last for weeks.

Smoking has been proven to cause most lung cancers.

Lung Cancer

More people die from lung cancer than any other type of cancer. Cancer occurs when cells of the body grow too quickly and don't work the way they should. Breathing becomes difficult for people with lung cancer. The cancer cells often spread to other organs as well. Although there are some treatments, especially when the disease is discovered early, many people still die.

DOCTORS DIAGNOSE

How do doctors diagnose a patient's lung problems? They'll need to get some information first. When did the problems start? Does anyone in their family have the same problem? Do they smoke?

Next, the doctor listens to the patient's chest with a **stethoscope** to make sure air is moving in all areas of the lungs. The doctor also listens for extra sounds, such as the "wheezing" usually heard in asthma or the "crackling" sometimes heard in pneumonia. The doctor may also check other parts of the body for signs of disease, including the back of the throat, nasal passages, ears, and even the skin. New machines called pulse oximeters use light, usually shining on the patient's finger, to quickly and painlessly measure how much oxygen is in the patient's blood.

A doctor will put a stethoscope on your back and ask you to breathe in and out. That helps them hear any issues in the lungs.

FUN FACT

IN AN EMERGENCY, A DOCTOR WILL CHECK FOR ABC: IS THE **A**IRWAY OPEN? IS THE PATIENT **B**REATHING? IS THE HEART **C**IRCULATING BLOOD?

Sometimes doctors need to take a look inside the body to see what's going on. They may request an X-ray. During an X-ray, energy passes through a person's body and onto special film. Solid structures in the chest, such as the ribs and heart, block some of this energy, while the lungs, because they're empty except for air, don't. Thus, the film can show physicians what the inside of a patient's chest looks like.

Like an X-ray, a CT scan uses energy passing through the body to take a picture of the inside. The CT scan takes a three-dimensional picture, however, and shows many more details than an X-ray. Another tool is a bronchoscope. This long, thin tube has a camera on one end to view a patient's larynx, trachea, and bronchi.

FUN FACT

CT SCANS ARE MORE COSTLY AND TAKE LONGER THAN X-RAYS.

An X-ray can give a doctor a good idea about what might be going on inside a person's lungs.

Are the Lungs Working?

X-rays and CT images show how the lungs appear. Pulmonary function tests (PFTs) are designed to measure how well the lungs work. Patients breathe into a machine that measures how much and how fast air is blown out. These tests are valuable for diagnosing asthma and emphysema. By comparing PFT results over time, doctors can determine if a patient is getting worse or if a new medication is helping.

HEALTHY HABITS

How can you keep your respiratory system healthy? You can make sure that you're not inhaling harmful things. If you work or play in dusty, dirty, or smoky conditions, wear a special mask. This is often important in areas that get a lot of wildfire smoke. Make sure you don't inhale carbon monoxide—a harmful, colorless, odorless gas—by keeping a carbon monoxide detector in your home. The biggest thing you can do: never start smoking cigarettes! Avoid cigarette smoke whenever possible.

Your daily habits are important too. Exercising helps to keep your lungs strong. Aerobic exercises are those that make you breathe heavily. You can run, jog, or skate. This is like exercise for your lungs. It helps keep them healthy!

Swimming laps in a pool is an aerobic exercise that works the whole body.

GLOSSARY

antibiotic: A drug that can kill bacteria in the body.

bacteria: Tiny, single-celled organisms. Some cause diseases in humans.

capillary: An extremely small blood vessel that can allow gases and nutrients to pass through its thin walls.

cartilage: A tough but flexible tissue used to support some body structures, such as the throat, ears, and joints.

cell membrane: The thin wall of a cell that separates it from the outside surroundings.

circulatory system: The body system made up of the heart, blood vessels, and blood.

essential: Completely necessary.

immune system: The parts of the body that fight germs and keep it healthy.

mucus: A thick, sticky fluid that helps trap dust and other particles.

pulmonary: Having to do with the lungs.

stethoscope: A special instrument used by medical professionals that allows them to hear very soft sounds coming from the chest.

virus: A nonliving package of proteins and genes that can cause diseases in humans.

FOR MORE INFORMATION

Books

Crane, Cody. *Respiratory System*. New York, NY: Children's Press, 2024.

Hicks, Dwayne. *Your Amazing Lungs*. Buffalo, NY: PowerKids Press, 2023.

Hyde, Natalie. *Investigating the Respiratory System*. St. Catharines, ON: Crabtree Publishing, 2023.

Websites

15 Facts About the Human Body!
www.natgeokids.com/uk/discover/science/general-science/15-facts-about-the-human-body/
Discover amazing facts about the human body with National Geographic Kids.

Lung
kids.britannica.com/kids/article/lung/353400
Explore more about the jobs and disorders of the lungs.

Quiz: Lungs & Respiratory System
kidshealth.org/en/kids/rsquiz.html
Take this quiz to see what you know about the respiratory system.

Publisher's note to educators and parents: Our editors have carefully reviewed these websites to ensure that they are suitable for students. Many websites change frequently, however, and we cannot guarantee that a site's future contents will continue to meet our high standards of quality and educational value. Be advised that students should be closely supervised whenever they access the internet.

INDEX